big & SMALL

Original Korean text by Yu-ri Kim
Illustrations by Hyeon-joo Lee
Korean edition © Aram Publishing

This English edition published by Big & Small in 2015
by arrangement with Aram Publishing
English text edited by Joy Cowley
English edition © Big & Small 2015

Distributed in the United States and Canada by
Lerner Publishing Group, Inc.
241 First Avenue North
Minneapolis, MN 55401 U.S. A.
www.lernerbooks.com

ISBN: 978-1-925186-16-1

Printed in Korea

FOSSILS TELL STORIES

Written by Yu-ri Kim

Illustrated by Hyeon-joo Lee

Edited by Joy Cowley

Brush! Brush!

Scrape! Scrape!

4

Dig, dig, dig!

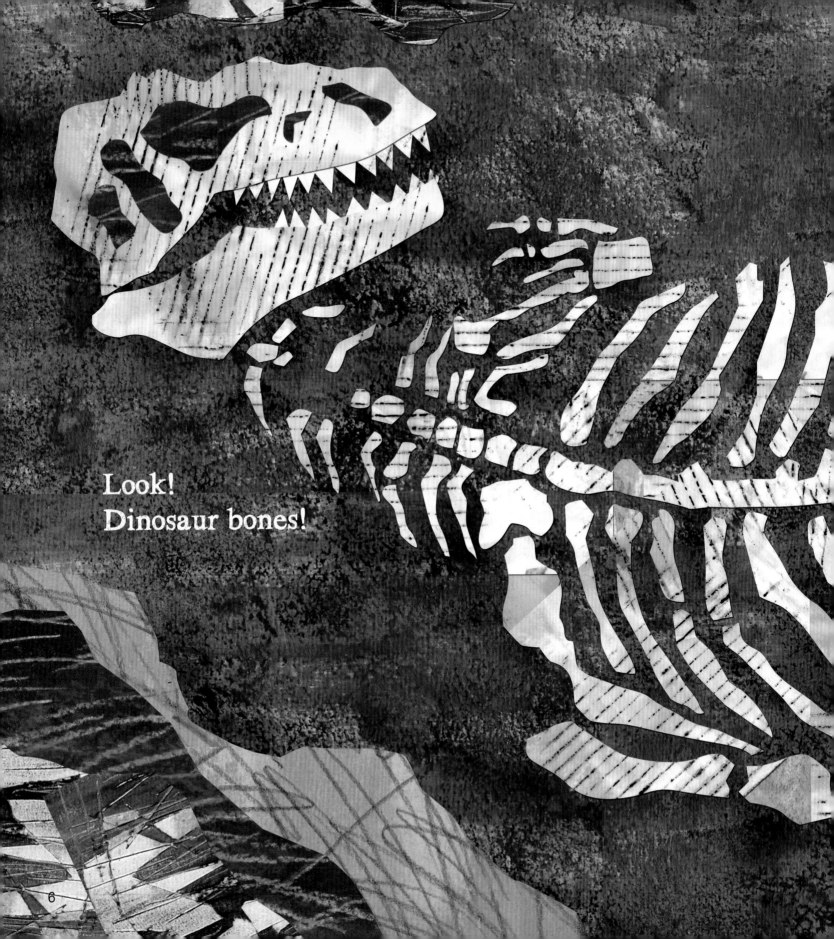

Look!
Dinosaur bones!

These are the fossils
of a dinosaur called Allosaurus.
From head to tail, it is as long
as the height of a four-story building.

Fossils are the remains of animals and plants buried and hardened underground, long ago.
Traces of animals and plants are also called fossils.

Did all dinosaurs turn into fossils when they died?

No. Just because they died
doesn't mean they became fossils.
Other animals could eat them.
The bones left over could decay
and disappear without a trace.

Then how did dinosaurs become fossils?

12

There may have been a landslide.
Something happened and dinosaurs
got buried under the earth.
Their flesh decayed
and only bones were left.

My dog buries bones in the earth.
Are they also fossils?

No, they are not fossils.
To turn into a fossil,
a bone has to be buried
for a long, long time
so that it becomes hard
like a rock.

What about clay pots
made by early people?

Those aren't fossils, either.
Things made by early people
are called artifacts.

Artifacts are things left
behind by people who
lived long ago.

17

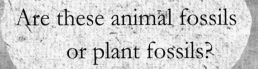

Are these animal fossils or plant fossils?

Fish fossil

If they look like the animals of today, then they are animal fossils.

Sea urchin fossil

Crab fossil

18

Tree leaf fossil

If they look like the plants of today,
then they are plant fossils.

Equisetum fossil

Sphenopsid fossil

Fossils of dinosaur footprints

The remains of animal footprints
or their crawl marks
are called fossils.

Dinosaur skin fossil

Dinosaur dung fossil

Dinosaur eggs fossil

Remains left by living creatures are also fossils.

Reptile dung fossils

By looking at fossils, we can work out
what kind of creatures lived long ago.

Ammonite
A sea animal with a
wrinkled coiled shell

Cephalaspis
A fish with a head
shaped like a shield

23

Fossils tell us how animals lived a long time ago.

It must have been a ferocious hunter.

Allosaurus

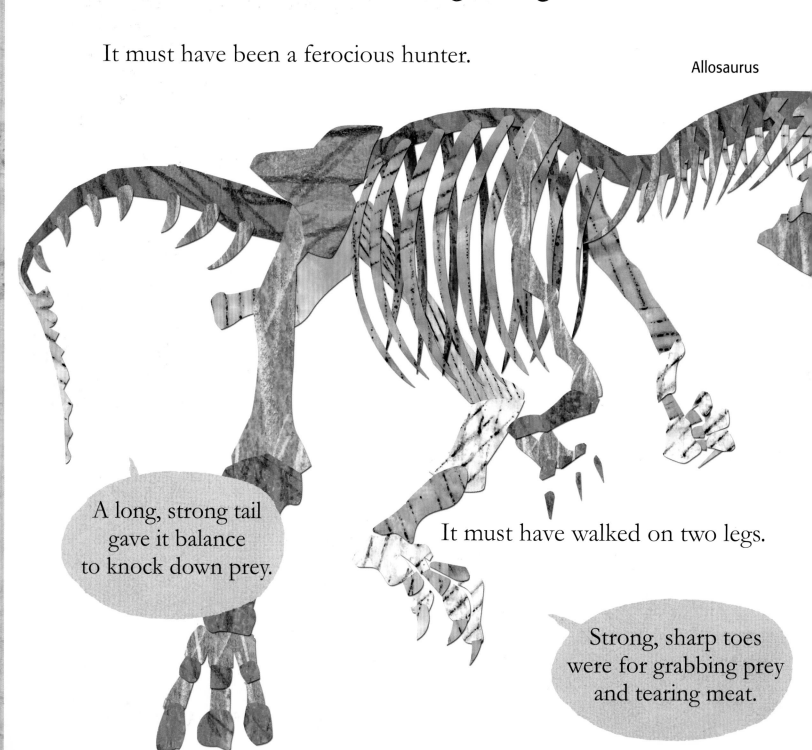

A long, strong tail gave it balance to knock down prey.

It must have walked on two legs.

Strong, sharp toes were for grabbing prey and tearing meat.

It must have had very hard armor.

It must have been a carnivore.

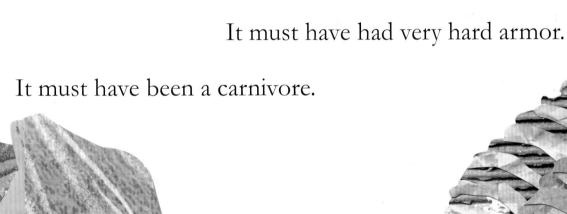

Trilobite

Large, sharp teeth
were for chewing meat.

When an enemy came,
it protected itself
by curling up.

It must have lived on water and on land.

Tiktaalik

Fins had leg bones
for swimming
and walking.

Fossils tell us stories
about ancient lives.
We can figure out many things
by looking at these bones.

27

Roar!

How Are Fossils Made?

1. Fish and clams live under the sea.

2. When fish die, they sink down.

3. The dead fish are buried under layers of mud.

Let's Make a Fossil!

1. Pick a leaf off a tree.

2. Put the leaf on a flat piece of soft clay.

3. Put another piece of soft clay over the leaf.

4. Their flesh decays but the hard parts of their bodies remain.

5. Over a long time, the remains become as hard as rock.

6. More time passes. The earth cracks and the fossils are revealed.

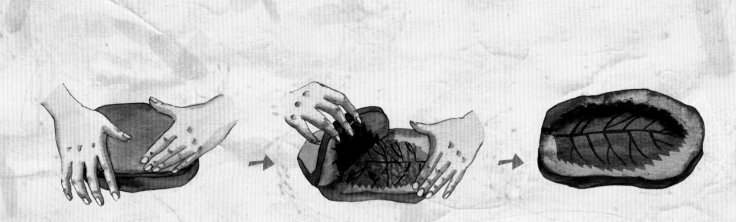

4. Press evenly all over.

5. Remove the top layer of soft clay.

6. Dry the leaf-imprinted clay in the shade.

Fossils Tell Stories

Fossils are remains and traces of animals and plants that lived long ago.

Let's think

How do fossils form?

What do fossils tell us?

Where can we find fossils?

What is the difference between a fossil and an artifact?

Let's do!

Let's make a fossil!

Pick a leaf off a tree and place it on a flat piece of soft clay.

Put another flat piece of soft clay over the leaf.

Press evenly all over and carefully remove the top piece of clay.

Dry the leaf-imprinted clay in the shade.